Júlio Alves

Strauss Type Deep Foundations

Júlio Alves

Strauss Type Deep Foundations

Case Studies

Imprint
Any brand names and product names mentioned in this book are subject to trademark, brand or patent protection and are trademarks or registered trademarks of their respective holders. The use of brand names, product names, common names, trade names, product descriptions etc. even without a particular marking in this work is in no way to be construed to mean that such names may be regarded as unrestricted in respect of trademark and brand protection legislation and could thus be used by anyone.

Cover image: www.ingimage.com

This book is a translation from the original published under ISBN 978-613-9-60427-2.

Publisher:
Sciencia Scripts
is a trademark of
Dodo Books Indian Ocean Ltd. and OmniScriptum S.R.L publishing group

120 High Road, East Finchley, London, N2 9ED, United Kingdom
Str. Armeneasca 28/1, office 1, Chisinau MD-2012, Republic of Moldova, Europe
Printed at: see last page
ISBN: 978-620-7-27874-9

ACKNOWLEDGEMENTS

To God for giving me the strength, perseverance and courage to enthusiastically take on yet another learning opportunity and achievement.

In particular, I would like to thank my mother, Jacqueline Elaine. For believing in me and dedicating herself every day to making me a better human being. For being tolerant and an educator. For being my role model as a professional and as a person.

To my brother and stepfather for their encouragement, affection and support.

To Santa Rita College - FASAR for making it possible to take this course.

To the supervisor of this work, Professor Roberta, for her dedication, patience and always reliable and efficient help.

To engineer Magno, for sharing his knowledge.

Finally, I would like to thank all my family and friends who have always cheered me on and believed in me, and who have contributed in some way to my learning.

EPIGRAPH

"Success is born of desire, determination and persistence in reaching a goal. Even if you don't reach the target, those who seek it out and overcome obstacles will at the very least do admirable things."

José de Alencar

SUMMARY

The aim of this paper is to present the concepts and definitions of the main types of deep foundations and their implications for civil engineering, with reference to the requirements of current standards. Subsequently, the case study of the Strauss pile construction process, located in the municipality of Barbacena/MG, will be presented. We will then highlight the characteristics that make the use of Strauss piling feasible and its concepts, presenting the executive method that was technically monitored through visual inspections. Based on this, the study analyses the entire Strauss pile construction process, discusses the problems of execution errors based on the prescriptions of technical standards and presents systematic and rational solutions for its construction process. The results analysed point to a significant improvement in the productive and economic aspects of the execution stage, highlighting the importance of good supervision of the execution of Strauss piles, as well as the importance of the design, execution and supervision standards for Strauss-type deep foundations.

Keywords: Deep foundations, civil engineering, visual inspection, Strauss pile.

SUMMARY

CHAPTER 1

INTRODUCTION

Many times in the construction of buildings and structures in general, problems associated with design errors and execution errors due to a lack of experience are encountered, making it necessary to alter the design and, in some cases, reinforce the foundations. The solutions used in these cases vary greatly depending on the type of problem, the demands made on the structure, the physical space available to carry out these alterations or reinforcements and the costs of each solution.

According to Rebello (2008), the problems that occur in foundations can come from natural phenomena or design and execution errors. Natural phenomena are, in principle, unpredictable, as they may be unknown until the foundation is laid. These problems are often localised and typical of a certain region. The problems resulting from the project are: errors in the load survey; errors in the design of the project, resulting from the choice of an inappropriate structural model; and the resistance of the soil, as a result of misinterpretation of the borehole.

This paper deals with a real case where several errors occurred due to a lack of experience in the execution of a Strauss pile for the foundation of a building.

1.1 OBJECTIVES

1.1.1 General objective

The aim of this work is to study the types of deep foundations most commonly used in civil construction, with an emphasis on the Strauss type.

1.1.2 Specific objectives

- Present a literature review on the main types of foundations in civil construction, and characterise the concepts and definitions of the main deep foundations.
- Through the case study, the details of the Strauss type foundation are shown, showing the

entire executive process that illustrates its use and its advantages and disadvantages.

- Characterise the problems related to the Strauss pile construction process, based on the requirements of technical standards, and present viable and applicable solutions to the problems encountered during construction.
- Draw up a cost estimate for the proposed foundation, comparing the costs before the execution errors foreseen in the project, and after the errors.

1.2 Justification

The purpose of this work was to study the execution of Strauss piles for the construction of a residential building in Barbacena / MG. Because of this, the execution of the Strauss pile must be supervised and follow the technical standards so that there are no errors and damage to the work.

CHAPTER 2

THEORETICAL FRAMEWORK

2.1 Geotechnical investigations

According to Rebello (2008), the idea of the soil's physical properties is very important, but so is the choice of the type of foundation and its sizing, as well as the determination of accidents such as the existence of water, cliffs and voids that could affect the construction process itself.

> Boring is a procedure that aims to get to know the natural conditions of the soil, in order to recognise its type, physical characteristics and especially its resistance. It also makes it possible to determine the depth of the water table (water in the subsoil) (REBELLO, 2008, p.27).

According to Quaresma (1998), in practice there is a predominance of field trials:

- The Standard Penetration Test (SPT);
- The rotary drilling test;
- The cone penetration test (CPT);
- The cone penetration test with measurement of neutral pressures or piezocone (CPT-U);
- The Vane Test;
- Pressure gauges (PMT);
- The Marchetti dilatometer (DMT);
- Plate loading tests or load tests;
- And geophysical tests such as the Cross-Hole test.

2.1.1 Standard Penetration Test (SPT)

The SPT (Standard Penetration Test) is the most widely used test in most countries around the world, and also in Brazil. In special cases where a more detailed analysis of the terrain is deemed appropriate, the CPT and CPT-U tests can be used.

According to Quaresma (1998), percussion drilling is a geotechnical field procedure capable of sampling the subsoil. When combined with the dynamic penetration test (SPT), it measures the resistance of the soil along the drilled depth. When carrying out a borehole, the aim is to find out:

- The type of soil crossed by taking a deformed sample every metre drilled;
- The resistance (N) offered by the soil to the standard sampler for each metre drilled;
- The position of the water level or levels, when encountered during drilling.

According to Rebello (2008), the SPT is an internationally standardised drilling process, so its results can be interpreted by anyone familiar with the method. The survey is carried out using equipment made up of a tripod, which actually has four legs, from which a standard weight of 65 kgf is dropped from a standard height of 75 cm. The weight drives a standardised steel tube into the ground, which is called a Terzaghi sampler. This sampler has an external diameter of 2" and an internal diameter of 1 3'8". The sampler is attached to a 1" rod which is screwed together as the sampler goes deeper into the ground.

According to Velloso and Lopes (2010), percussion boreholes are boreholes capable of rising above the water level and passing through relatively compact or hard soils. The hole is cased if it is unstable; if it is stable, drilling can continue without casing, possibly adding a little bentonite to the water. Drilling progresses as the soil, broken up with the aid of a core drilling rig, is removed by circulating water. The drilling equipment is shown in figure 1.

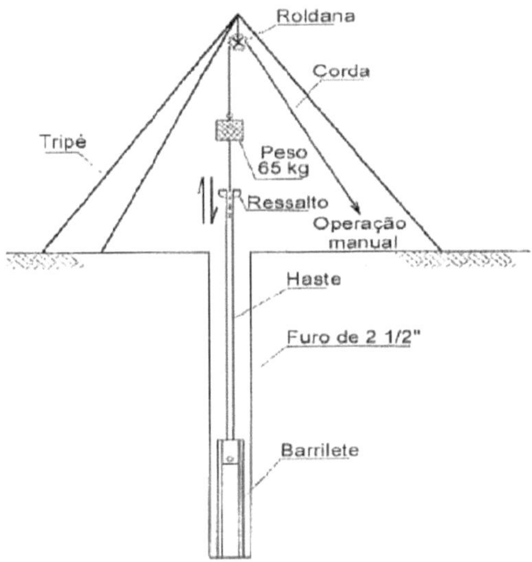

Figure 1: SPT test equipment
Source: Velloso and Lopes, 2010.

Advantages and disadvantages of using SPT, according to Velloso and Lopes (2010):

- Advantage: simple and economical;
- Disadvantage: prone to gross errors due to lack of maintenance of the sampler or deficiencies on the part of the technician.

According to ABEF (2012), the states of compactness and consistency are estimated according to the penetration resistance index (N), as shown in Table 1, sandy silts are classified by compactness and clayey silts by consistency.

Table 1: States of compactness and consistency

COMPACTNESS AND CONSISTENCIES ACCORDING TO PENETRATION RESISTANCE S.P.T.		
SOIL	NAME	NO. OF BLOWS
Compactness of sand and sandy silt	Cute	<4
	Not very compact	5 - S
	Compact measurement	9 - IS
	Compact	19-41
	Very compact	>41
Consistency of clays and clayey silts	Too soft	<2
	Soft	2-5
	Average	6 - 10
	Rija	1119
	Hard	> 19

Téchne (2004) mentions that the final report includes the location plan, the situation and the RN (Level Reference) of the boreholes, descriptions of the soil layers, the penetration resistance index, the resistance x depth graph, the macroscopic classification of the layers, the depth and limit of percussion drilling per borehole and also the existence or not of a water table and the initial level after 24 hours.

2.2 Foundations

According to Velloso and Lopes (2010), the foundation is the structure responsible for transmitting loads from the building to the ground. It is the first part of the building to be constructed and is in direct contact with the ground.

According to ABCP (2002), the foundation system consists of the building's structural element below ground, which can be a block, pile or pipe, and the surrounding soil mass under the base and along the shaft. Its function is to safely support the loads coming from the building.

According to Velloso and Lopes (2010), the main items to consider in order to have a safe foundation that fulfils its intended role are: the

elements need strength to withstand the stresses generated by the stresses, as well as a rigid soil that provides support without excessive deformation and settlement.

According to Velloso and Lopes (2010), the elements needed to develop a foundation project are:
- Topography of the area

- Topographic survey (planialtimetric);
- Data on embankments and slopes on the ground (or that could reach the ground).

- Geotechnical data
- Subsoil investigation (sometimes in two stages: preliminary and complementary);
- Other geological and geotechnical data (maps, aerial and satellite photos, aerophotogrammetric surveys, articles on previous experiences in the area).

- Data on neighbouring buildings

- Number of floors, average load per floor;
- Type of structure and foundations;
- Foundation performance;
- Existence of subsoil.
- Possible consequences of excavations and vibrations caused by the new work.

- Data on the structure to be built

- Type and use of the new building;
- Structural system (hyperstaticity, flexibility);
- Construction system (conventional or precast);
- Loads (actions on foundations).

2.3 Classification of foundations

According to NBR 6122 (2010), foundations are divided into two categories:

- Shallow foundations;
- Deep foundations.

According to Velloso and Lopes (2010), the distinction between these two types is made according to the (arbitrary) criterion that a deep foundation is one whose base failure mechanism does not appear on the surface of the ground. As the base failure mechanisms above it typically reach twice their smallest dimension. NBR 6122 (2010) determined that deep foundations are those whose bases are implanted at a depth greater than twice their smallest dimension, and at least 3 metres deep. Figure 2 shows a diagram of shallow and deep foundations.

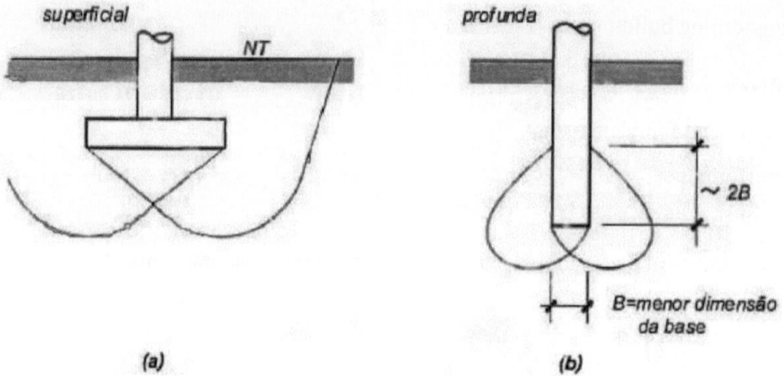

Figure 2: Schematic of shallow and deep foundations

Source: Velloso and Lopes, 2010.

2.4 Shallow foundations

According to Rebello (2008), a superficial or direct foundation is defined as one in which the building's loads (superstructure) are transmitted to the soil in the very first layers. For this to happen, it is obviously necessary for the soil in these first layers to have sufficient resistance to withstand these loads. Deciding on the type of foundation requires knowledge of the soil, provided by drilling.

According to Rodrigues (2006), in a more practical way, he describes a direct or shallow foundation as: "the fact that the load distribution from the pillar to the soil occurs through the base of the foundation element, whereby the approximately point load that occurs on the pillar is transformed into a distributed load, to such an extent that the soil is capable of bearing it".

According to NBR 6122 (2010), the following are examples of direct foundations: block, footing and radier, as shown in Table 2.

Table 2: Characteristics of shallow foundations

Types	When to use	Cost	Executive Characteristics
Block	Used when the soil is highly resistant, with no restrictions on high loads	Low	Simple execution
Sapa ta		Low, but larger than the block for low loads	Simple execution It can take different geometric shapes to facilitate the support of eccentrically shaped pillars.
Radier	When footings come close to each other or overlap	High cost	High deadline, due to the need to clear the entire area before starting the work.

12

	When you want to standardise settlement

Source: MANUAL DE ESTRUTURAS ABCP, 2002.

2.4.1 Block

According to ABCP (2002), it is a simple concrete foundation element, designed so that the tensile stresses produced in it can be resisted by the concrete, without the need for reinforcement; figure 3 shows a concrete block.

Figure 3: Concrete blocks

Source: Available at: < http://esporte.ig.com.br/futebol/em-meio-a-polemicas-wtorre-inicia-colocacao-de-pilares-da-arena/n1597349546237.html> Accessed on: 09 November 2017.

2.4.2 Shoes

According to ABCP (2002), it is a reinforced concrete foundation element, smaller in height than the block, using reinforcement to resist tensile forces.

According to TÉCHNE (2004), the main types of footings are:

Isolated: they only receive loads from one column. They are generally the most economical solution because they consume less concrete. Footings can have various shapes, but the most common is the conical rectangular one, as it consumes less concrete and requires simpler formwork. In the case of non-rectangular columns, the footing should have its centre of gravity coinciding with the centre of load, as shown in figure 4.

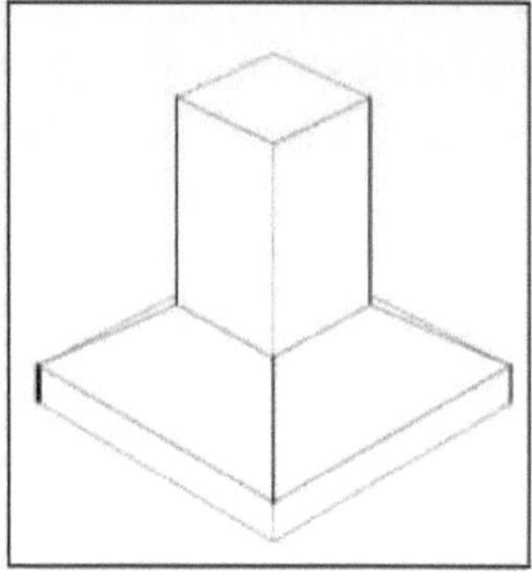

Figure 4: Diagram of an insulated footing
Source: TÉCHNE, 2004.

Associated: used when there are columns very close together and the isolated footings would overlap. They may also be necessary when the structural loads are high. As with isolated footings, the positioning of the foundation piece must respect the load centre of the columns, according to the diagram in figure 5.

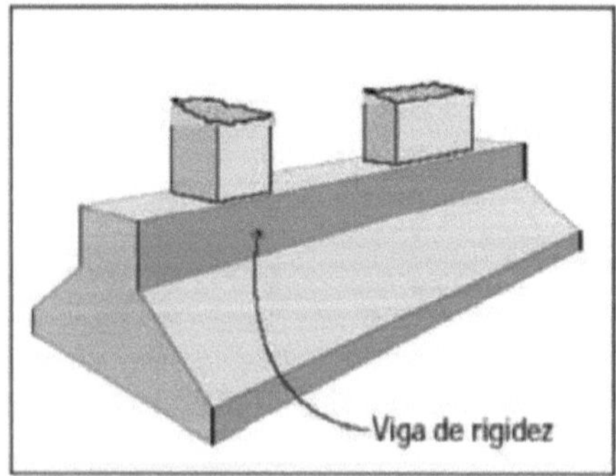

Figure 5: Schematic of associated footing

Source: TÉCHNE, 2004.

Leveraged: if the project foresees a footing on a land border or with an obstacle, the part cannot have the centre of gravity of the footings and the load centre of the columns coincide. To compensate for the eccentricity of the loads, it is necessary to transfer part of the forces to a nearby footing by means of a levered beam, as shown in figure 6.

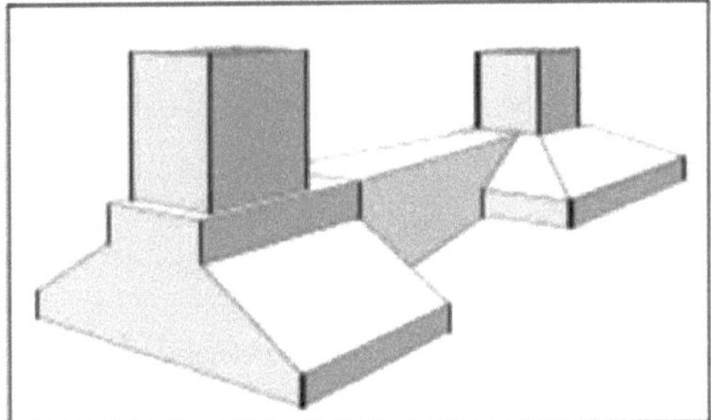

Figure 6: Leveraged shoe diagram

Source: TÉCHNE, 2004.

Corridors: receive loads directly from the walls. The load is transferred linearly. Running footings are a substitute for foundations, for more heavily loaded walls or less resistant soils, according to the diagram in figure 7.

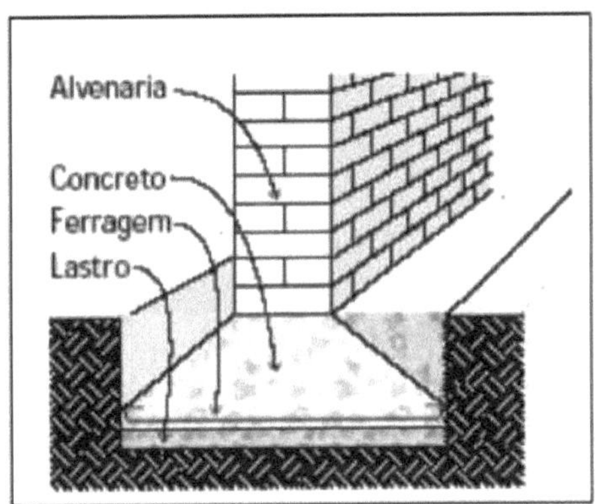

Figure 7: Schematic of a slipform foundation

Source: TÉCHNE, 2004.

2.4.3 Radier

According to Teixeira and Godoy (1998), when all the pillars of a structure transmit their loads to the ground through a single footing, this is called a radier foundation.

According to NBR 6122 (2010), the term radier should only be used when an associated surface foundation receives all the columns of the building (general radier), or when it receives only part of the columns of the building (partial radier). From a design point of view, however, these two cases can be treated in the same way.

According to Velloso and Lopes (2010), a radier foundation is adopted when:

- the areas of the footings come closer to each other, or even interpenetrate (as a result of high loads on the pillars and/or low working stresses);
- you want to even out the settlements (through an associated foundation).

In terms of shape or structural system, radials are designed according to four main types: flat radials, radials with pedestals or mushrooms, ribbed radials and box radials, as shown in figure 8.

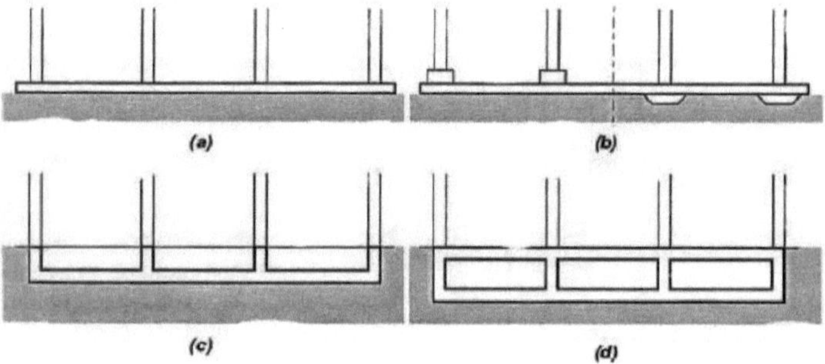

Figure 8: Radiers: (a) plain; (b) with pedestals or mushroom slabs; (c) ribbed (inverted beams) and (d) boxed

Source: Velloso and Lopes, 2010.

2.5 Deep foundations

According to NBR 6122 (2010), deep foundations are those that receive the load from the superstructure and discharge it into the ground through the base (tip resistance) or lateral surface (stem resistance) or a combination of the two. The elevation must be more than twice its smallest dimension in plan, and at least 3.0 metres. Deep foundations are considered to be piles or tubulars.

According to Budhu (2015), deep foundations are used when:

- the near-surface soil does not have sufficient load-bearing capacity to support the structural loads of shallow foundations;

- the estimated soil settlement exceeds the permissible limits, i.e. the settlement is greater than the serviceability limit state;

- differential settlement due to soil variability and uneven structural loads is excessive;

- structural loads consist of lateral loads or lifting forces;

- excavations to build a shallow foundation on a competent soil layer are difficult and costly.

2.5.1 Pipelines

According to Alonso and Golombek (1998), tubulars are structural elements of deep foundations built by concreting an open pit (lined or not) in the ground, usually with an enlarged base. They differ from piles in that, at least in their final stage, labourers descend to complete the geometry of the excavation or clean the soil. Pipelines are divided into two basic types: open-cast (usually unlined) and compressed-air (or pneumatic), which are always lined, the lining being either a reinforced concrete jacket or a steel (metallic) jacket. In this case, the metal jacket can be lost or recovered.

According to Rebello (2008), a tubular foundation is a deep foundation made up of a vertical concrete cylinder, which may or may not be widened at the base. The vertical cylinder, known as the shaft, is built like a well. The minimum diameter of the shaft is 70 cm, to allow one labourer to work on it when it is done manually. When built mechanically, the diameter of the shaft is adapted to both the loads to be borne and the dimensions of the equipment. The base can be circular or elongated (false ellipse) and its height must be limited to 2 metres, as shown in figure 9. Ultimately, the base is a footing made at a great depth. Its function is to transmit the load of the superstructure to the ground, distributing it over its area.

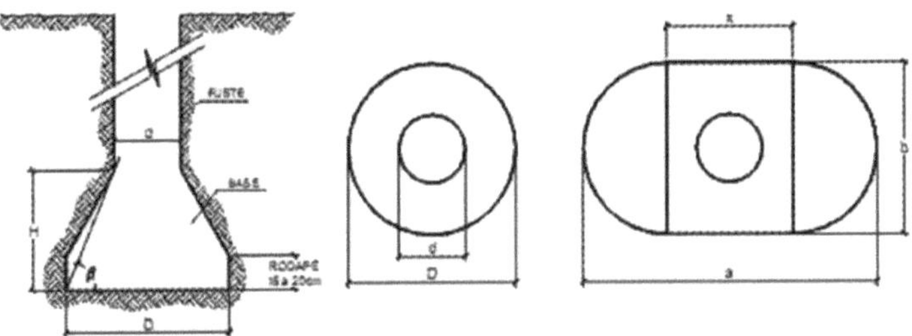

Figure 9: Profile, circular and false ellipse pipework

Source: Available at: http://construcaociviltips.blogspot.com.br/2011/07/tubuloes.html

Accessed on: 09 November 2017.

- Open-cast pipelines

According to NBR 6122 (2010), in this type of foundation the loads are essentially transmitted by the base to a more resistant substrate. This type of foundation is used above the water table, or even below it, in cases where the soil is more resistant.

It is also possible to control the water inside the pipe, while respecting safety regulations. Figure 10 shows a schematic of an open-cut pipe.

Figure 10: Schematic of an open-cut pipeline

Source: Available at: https://www.escolaengenharia.com.br/tubulao-a-ceu-aberto/
Accessed on: 09 November 2017.

• Compressed air pipelines

According to NBR 6122 (2010), this type of solution is used whenever the aim is to construct pipework below the water level in soils that do not fulfil the conditions. The excavation of the shaft of these pipes is always carried out with the aid of a lining, which can be made of concrete or steel (lost or recovered). Figure 11 shows a diagram of a compressed air pipeline.

Figure 11: Compressed air pipework

Source: Available at: http://rocafundacoes.com.br/tubuloes-sobre-ar-comprimido/index.html Accessed on: 09 November 2017.

According to Rebello (2008), tubular foundations are preferably used for large-scale works, mainly bridges and viaducts. However, in special situations, it can be used for smaller loads, such as foundations in difficult terrain to which other equipment cannot access.

2.5.2 Stakes

According to ABCP (2002), a pile is a foundation element that is executed with the aid of tools or equipment, which can be driven by percussion, pressing, vibration or excavation.

According to Velloso and Lopes (2010), foundations can be classified according to different criteria. Depending on the material, they can be made of wood, concrete, steel or mixed materials. According to the execution process, piles can be separated according to the effect they have on the soil and are classified as driven and excavated.

- Driven piles

Displacement, where driven piles would generally be, since the soil in the space that the pile will occupy is displaced horizontally.

According to Velloso and Lopes (2010), piles driven into granular soils that are slightly to moderately compact cause densification or an increase in the compactness of these soils to the extent that the volume of the pile introduced into the ground leads to a reduction in the void ratio. This effect is beneficial from the point of view of the pile's behaviour (a higher load capacity and lower settlements are obtained than if the soil were kept in its original state). If the soil is already very compact, the introduction of the pile will no longer cause an increase in compactness.

- Excavated piles

Replacement, where excavated piles in general would be, since the soil in the space that the pile will occupy is removed, causing some level of reduction in horizontal geostatic stresses.

Velloso and Lopes (2010) mention that excavated piles can cause decompression of the ground, which will be greater or lesser depending on the type of support. At one extreme would be piles excavated without support (which is only possible in soils with a certain percentage of fines and above water level), where decompression is pronounced. At the other extreme would be piles excavated with the aid of metal jackets, which advance at practically the same level as the excavation tool, where there is very little relief. In the middle of these extremes would be piles excavated with the aid of stabilising fluid or mud.

According to Velloso and Lopes (2010), Table 3 shows all types of piles, indicating whether they are displacement or replacement piles.

Table 3: Pile types

TYPE OF EXECUTION	STAKES
Travelling	Madeira
	Metallic
	Precast concrete
	Mega
	Franki
Replacement	Root
	Continuous propeller **type** in general
	With bentonite mud
	Strauss

Source: Adapted from Velloso and Lopes, 2010.

2.5.3 Wooden stake

According to Velloso and Lopes (2010), wooden piles are nothing more than tree trunks, as straight as possible, usually driven by percussion using free-falling pylons. If they are to be used in permanent works, they must be treated with preservative products.

According to Alonso (1998), eucalyptus is the most widely used wood in Brazil, mainly as a foundation for temporary works.

There are the following requirements for timber piles in the Brazilian standard NBR 6122 (2010):

• The tip and top must have diameters greater than 15 and 25 cm respectively, and a straight line segment connecting the centres of the tip and top sections must be entirely inside the pile;
• The tops of the piles must be protected by suitable shock absorbers to minimise damage during driving. If any damage occurs to the pile head during driving, the affected part must be cut off. When penetrating or passing through resistant layers, the tips should be protected by steel spikes, as shown in figure 12. As for splices, they can be made by splicing, by splices or by a metal ring, as shown in figure 13.

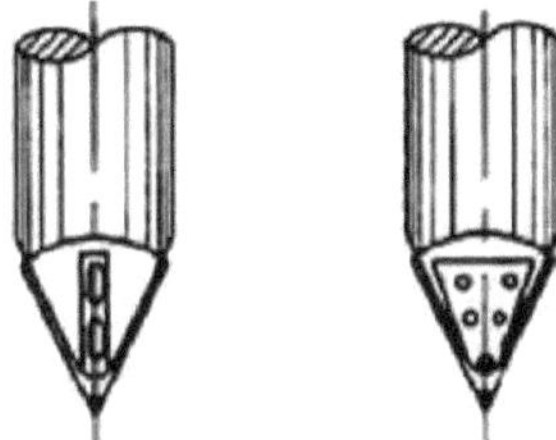

Figure 12: Tip reinforcement

Source: Alonso, 1998.

Figure 13: Types of splice

Source: Alonso, 1998.

According to Alonso (1998), the admissible load of wooden piles, from a structural point of view, depends on the diameter of the pile's middle section, as well as the type of wood used. However, the values shown in Table 4 are usually adopted as the order of magnitude:

Table 4: Permissible loads normally used in timber piles

Diameter (cm)	Load (kN')
20	150
25	200
30	300
35	400
40	500

Source: Alonso, 1998.

Azeredo (1997) presents some advantages and disadvantages of wooden piles.

- Advantages:

- no problems with transport and handling;

- easy cut;

- easy to obtain in variable lengths;

- easy splice;

- low cost.

- Disadvantages:

- low durability;

- wood is subject to rot, caused by an aerobic fungus whose development depends on the coexistence of air and water. Wooden piles must therefore always be submerged. A change in the water level will weaken the transition zone between the water level and the air;

- he average lifespan of a wooden pile when the water table is lowered is 8 to 10 years.

2.5.4 Metal piling

According to Velloso and Lopes (2010), steel piles come in a variety of forms, from profiles (rolled or welded) to tubes (calendered and welded or seamless). Rolled profiles include rails, which are generally used after they have been removed from the railway (used rails). The profiles can be used alone or in combination (double or triple), figure 14 shows all the profiles used.

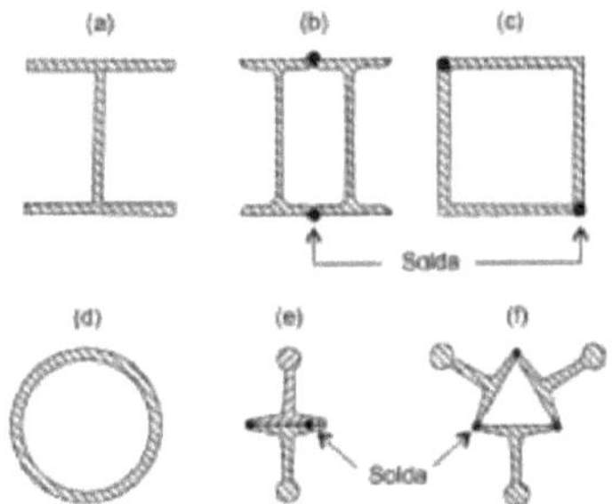

Figure 14: Steel piles (cross-sections): (a) welded plate profile; (b) rolled I profiles, double associated; (c) angle profiles; (d) tubes; (e) double associated rails; (f) triple associated rails.

Source: Velloso and Lopes, 2010.

According to Alonso (1998), nowadays the problem of corrosion of metal piles when they remain entirely buried in natural soil is no longer questioned, because the amount of oxygen that occurs in natural soils is so small that the chemical reaction, as soon as it starts, completely exhausts this component responsible for corrosion, as shown in figure 15.

Figure 15: Metal pile

Source: Available at: https://sites.google.com/site/fundacoesecvufsc/visitasobra2

Accessed on: 09 November 2017.

Azeredo (1997) presents some advantages and disadvantages of metal piling.

- Advantages:

- easy handling and transport;
- easy to drive into the ground due to the reduced thickness of the sheets;
- are obtained at any length, without any loss whatsoever;
- easy to cut and splice using electric welding;

- Disadvantages:

- high cost.
- The behaviour of a steel pile in aggressive water is not very dangerous, as long as the water is not moving; in still water, the resulting layer of rust is enough to protect the pile, while in running water the oxidised surface is carried away, exposing the pile to another attack, and so on;

2.5.5 Precast concrete piles

According to Alonso (1998), concrete piles are one of the most suitable construction materials for making piles, particularly pre-cast piles, due to the quality control that can be exercised during both the making and driving process. They can be made from reinforced or prestressed concrete that is densified by centrifugation or vibration. Figure 16 shows the cross-sections of precast concrete piles.

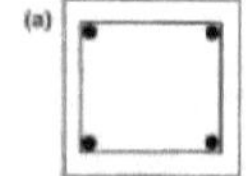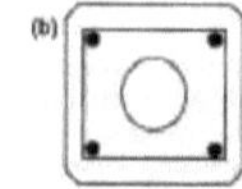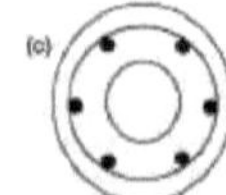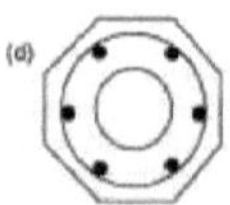

Figure 16: Cross sections

Source: Velloso and Lopes, 2010.

According to Azeredo (1997), precast concrete piles have some advantages and disadvantages:

24

- Advantages: they have good load-bearing capacity and good resistance to bending and shear stresses. In addition, because they are produced in appropriate factories, they have good concrete quality and are controlled and inspected by laboratories.
- Disadvantages: Because they are made of reinforced or prestressed concrete, they have a high self-weight, limiting sections and lengths due to transport, and cuts and splices are difficult to make.

According to Velloso and Lopes (2010), prestressed concrete piles are used for large loads and long lengths and have the following advantages:

- high compressive strength;
- greater handling and crimping capacity;
- ability to withstand high tensile forces;
- easy to mould with any cross-section configuration.

According to Azeredo (1997), a problem with precast piles arises when they are driven in the presence of aggressive water. The water, penetrating through the concrete, will reach the reinforcing irons, which, as they oxidise, increase in volume, bursting the concrete and leaving the reinforcement even more exposed to the aggressiveness of the water.

Figure 17 shows the machinery driving the precast concrete piles.

Figure 17: Precast concrete piles

Source: Available at: http://www.tecgeo.com.br/servicos/estacas-pr-moldadas-de-concrete-3 Accessed on: 09 November 2017.

According to Velloso and Lopes (2010), the following resources are used to protect the pile:

a) calculation of the concrete in stage I, which avoids the appearance of cracks during lifting;
b) painting the pile with asphalt-based products;
c) vitrification of the pile, which, however, reduces the desired elasticity for lifting the pile.

2. 5.6 Mega pile

For Azeredo (1997), mega piles are pre-moulded concrete elements that are driven by pressing using a hydraulic jack. They are used to reinforce foundations or replace existing foundations, using the structure itself as a reaction.

- Advantages: They are indicated for the recovery of pathologies without the use of demolition.

- Disadvantages: High cost and long driving time.

According to NBR 6122 (2010), driving in porous soils can be aided by saturating the soil and compacted sand with water jets from the inside of the segment. When the segments are made of concrete, the splice will be made by simple superimposition or by solidification specified in the project. Segments will be spliced using welds or threads. Figure 18 shows megas piles reinforcing a structural pile that was collapsing.

Figura 18:   Reinforcement pile

Source: Available at: http://www.basemreforco.com.br/ Accessed on: 09 November 2017.

2.5.7 Franki pile

According to ABCP (2002), a cast-in-place reinforced concrete pile uses a casing pipe with a closed end, so that there is no depth limitation due to the presence of underground water.

According to Maia (1998), the execution stages of the Franki pile are shown below, in figure 19 a schematic of these execution stages is shown.

Stage 1: Piling begins with positioning the casing pipe and forming the dowel;

Stage 2: After supporting the pipe on the ground, a certain amount of gravel and sand is thrown inside to be compacted by the impact of the pestle blows and to expand laterally, adhering strongly to the pipe;

Stage 3: The pipe is then driven into the ground by the impact of repeated blows from the pylon on the dowel. The final crimping depth is determined by checking the negativity of the pipe in the last few metres of crimping;

Stage 4: Once driving is complete, the pipe is attached to the pile driver's tower by means of steel cables, in order to expel the dowel and begin the execution of the widened base;

Stage 5: The widening of the base is achieved by heavily tamping down successive small quantities of almost dry concrete (zero slump);

Stage 6: Once the widened base is finished, the reinforcement is laid, then the shaft is concreted by pouring successive layers of concrete at a low height and recovering the pipe by tamping the layers. Concrete pouring of the shaft is completed about 30 cm above the levelling level.

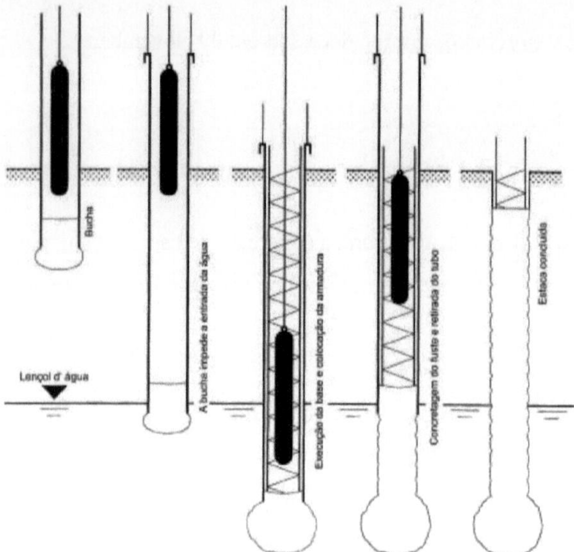

Figure 19: Standard Franki pile construction
Source: Velloso and Lopes, 2010.

According to Azeredo (1997), this pile has some advantages and disadvantages.

- Advantages:
- It is executed to the length that is strictly necessary (an advantage common to all *cast-in-place* piles*);*
- Great adherence to the ground due to the roughness of the stem;
- Better pressure distribution thanks to the wider base;
- High load capacity.

- Disadvantages:
- Setting of concrete in contact with the ground;
- The great vibration caused during driving can damage neighbouring buildings;

- The tensile rupture of uncured concrete or the loss of contact between the base and the supporting soil due to the lifting of an already driven pile, caused by the driving of neighbouring piles;

- Strangulation of the shaft when concreting through very soft soils with the damage caused by the shaft breaking during concrete tamping, caused by the sudden shortening of the reinforcement.

2.5.8 Root pile

According to NBR 6122 (2010), the root pile is characterised by its execution by rotary or rotopercussive drilling and by the use of a casing (a set of recoverable metal pipes) that is integral to the stretch of soil, and which is completed by placing a frame along its entire length and filling it with cement-sand mortar. The mortar is compacted using pressure, usually compressed air, as shown in figure 20.

Figure 20: Root pile being driven

Source: Available at: http://tecsonda.com/estaca-raiz/ Accessed on: 09 November 2010 2017.

According to Alonso (1998), they are those in which compressed air injections are applied immediately after moulding the stem and on top of it, at the same time as the coating is removed.

For Alonso (1998), the execution of the root pile is characterised by four consecutive phases, listed below:

- drilling aided by water circulation;

- reinforcement installation;
- filling with mortar;
- removing the coating and applying compressed air blows.

According to Velloso and Lopes (2010), there are some peculiarities of the root pile that allow it to be used in cases where other types of pile cannot be used:

- do not produce shocks or vibrations;
- There are tools that allow them to be carried out through obstacles such as blocks of rock or pieces of concrete;
- the equipment is generally small, which makes it possible to work in restricted environments;
- can be carried out vertically or at any inclination.

2.5.9 Continuous auger pile

According to NBR 6122 (2010), it is a *cast-in-place* concrete pile, made by rotating a continuous helical auger into the ground and injecting concrete through the auger's central shaft at the same time as it is removed. The reinforcement is always placed after the pile has been concreted.

According to Antunes and Tarozzo (1998), the equipment normally used to drive the auger into the ground consists of a metal tower, the height of which is appropriate to the depth of the pile, fitted with two guides at the ends, the lower guide of which can be replaced by an auger cleaner, a hydraulically driven rotary table with torque appropriate to the diameter and depth of the pile to be driven, and a winch compatible with the necessary starting efforts, as shown in figure 21.

Figure 21: Continuous auger pile in progress
Source: Available at: https://www.emaze.com/@AOOILQIQ Accessed on: 09 November 2017.

Velloso and Lopes (2010) point out that there is a technical discussion about the classification of continuous auger piles, whether they should be considered as traditional "replacement" excavated piles, or as "no displacement" piles. According to the executive process, if practically all the soil in the space where the pile is to be built is removed, it should be classified as a "replacement" pile. If, during the construction process, there is lateral displacement of the soil to create the space for the pile, it can be considered a "no displacement" pile or even a "small displacement" pile.

- With ground displacement

According to Velloso and Lopes (2010), at least two types of soil-displacing helical piles should be mentioned, because they differ in that the helical tool (or auger) that penetrates the ground is designed in such a way as to move the soil laterally when the tool is inserted or extracted.

- Omega piles: this pile can be made with diameters from 30 cm to 60 cm, and lengths of up to 35 metres. The permissible load can reach 2000 kN.

• Atlas pile: this type of pile can also be made in diameters of 36 to 60 cm, and can reach lengths of up to 25 metres. It is similar to the Omega pile, but differs in the way the pipe is removed, which is done by rotating it in the opposite direction to when it is inserted.

• With soil excavation

According to Velloso and Lopes (2010), this type of pile is made with a long propeller auger, made up of spiral plates that develop around the central tube. The lower end of the auger is fitted with claws to make it easier to cut through the ground, and a cover that prevents soil from entering the centre tube during excavation. The most common equipment allows piles to be driven with diameters of 30 cm to 100 cm and lengths of 15 m to 30 m.

Antunes and Tarozzo (1998) present some advantages and disadvantages of having a continuous auger pile:

• Advantages:
- the possibility of execution very close to the boundary, thus reducing the eccentricities between the loads of the pillars and the centre of the piles;

- high productivity significantly reduces the work schedule with just 1 labour team;
- is flexible on most types of terrain, except in the presence of boulders and rocks;
- the execution process does not produce the disturbances and vibrations typical of percussion equipment and does not cause decompression of the ground.

- Disadvantages:
- depending on the size of the equipment, work areas must be flat and easy to move around;
- Due to its high productivity, it requires a concrete plant close to the workplace;
- the need for a wheel loader on site to remove and clean up material extracted from drilling outside the work area;
- limitation on pile and frame lengths.

2.5.10 Excavated pile with bentonite slurry

According to NBR 6122 (2010), these are piles excavated using a stabilising fluid, which can be bentonite slurry to support the excavation walls. Concreting is submerged, with the concrete displacing the stabilising fluid upwards out of the tube. This type of pile has circular sections (called stations) and rectangular sections (called barrels).

According to Saes (1998), bentonite sludge consists of a mixture of water and bentonite. Bentonite is a clay from the montmorillonite family found in natural deposits. Its properties can vary greatly from one deposit to another.

Saes (1998) mentions that bentonite mud has three very important characteristics:

-) Stability, which means that the bentonite particles do not decant for a long period of time.
-) Ability to quickly form an impermeable film ("cake") on a porous surface (soil, filter paper).
-) Thixotropy, which consists of the reversible ability to become liquid when agitated or pumped and to gel (form a structure) when movement ceases.

According to Rebello (2008), the execution process for piles excavated with bentonite slurry comprises the following phases:
- simultaneous excavation and filling of the pile with previously prepared bentonite slurry;
- placing the previously assembled armour into the mud-filled excavation;
- pouring the concrete from the bottom up through concreting pipes (hopper pipes), which, being denser, expel the slurry, which is pumped back into the tanks.

Velloso and Lopes (2010) state that bentonite slurry must be used in the construction of excavated piles:
- contain the bottom and walls of the excavation by the action of hydrostatic pressure on them;
- easily dislodged and replaced by concrete;
- keep excavation waste in suspension, avoiding it being deposited at the bottom of the

excavation or in the pipework;

- easily pumpable.

Figure 22 shows the injection of the bentonite slurry into the metal jacket:

Figure 22: Detail of bentonite sludge

Source: Available at:

http://www.solonet.eng.br/Duvidas/estacas_estacao.htm Accessed

on: 09 November 2017.

According to NBR 6122 (2010), the bentonite slurry, once mixed, must be left to rest for twelve hours to fully hydrate and must have the characteristics shown in Table 5:

Table 5: Characteristics of bentonite sludge

Properties	Values	Test equipment
Density	1.025 g/cm^3 to 1.10 g/cm^3	Densimeter
Viscosity	30 s to 90 s	Funnel Marsh
pH	7a **11**	pH indicator
Sand content	**Up to** 3%	*Bar id sand contente or similar*

Source: NBR 6122, 2010.

2.5.11 Strauss pile

According to Velloso and Lopes (2010), it is a type of ground-moulded pile that requires relatively simple equipment: a tripod with a winch, a small pylon, an excavation tool and casing pipes. Its quality depends very much on the work of the team in charge.

According to NBR 6122 (2010), the integral coating ensures the stability of the borehole and guarantees the conditions so that the concrete does not mix with the soil or strangle the pile shaft. This type of pile should not be driven into submerged sands or very soft saturated clays.

According to the ABEF manual (2002), this pile requires great care in execution, especially when working below the water table, to avoid water entering the mould. This is achieved by placing references on the suspension cables, which make it possible to permanently assess the relative positions of the mould and the filling concrete.

According to the ABEF manual (2002), the equipment used to carry out Strauss pile services is listed below, and figure 23 shows the Strauss pile in progress.

- Wooden or steel tripod;
- Winch coupled to a diesel or electric motor;
- Mechanical percussion probe, made up of a tube of approximately 2.5 metres in diameter, smaller than the metal jacket, with a cutting shoe at the lower end, fitted with a hinged valve for removing soil at the lower end (also known as a piteira, or probe bucket);
- Solid, cylindrical metal socket or pestle weighing approximately 300kg;
- Steel casing pipes (metal jacket) 2.0 to 3.0 metres long, spliced with male and female threads;
- Crown (the first covering to be positioned, with its lower edge cut out in the shape of teeth);
- Manual winch to remove the pipe;
- Sheaves, wire ropes and tools;
- Concrete mixer (optional).

Figure 23: Strauss pile equipment in progress

Source: Available at: http://www.tecgeo.com.br/servicos/estacas-strauss-4

According to NBR 6122 (2010), all Strauss:

- Drilling:

The equipment must be positioned to ensure that the pile is centred and vertical. Drilling begins by applying repeated blows with the pestle or piteira to form a pre-drill at a depth of 1.0m to 2.0m, into which a short segment of casing with a crown at the tip is placed. Drilling then continues with repeated blows from the probe, which removes the soil. As the hole is formed, the casing pipes are inserted until the required depth is reached. Once drilling is complete, water is poured into the pipes to clean them. The water and mud are completely removed by the pit and the socket is washed.

- Concreting:

The concrete is poured through a funnel into the lining, in sufficient quantity to create a column of approximately 1.0 metre, which must be rammed to form the tip of the pile. As the pile continues to be driven, the concrete is poured and tamped down while the lining is removed. The dosage of the concrete must be such that a characteristic compressive strength of at least 20 per cent is obtained.

MPa, with a minimum cement consumption of 300 kg/m^3 must have a minimum slump of 8 cm for unreinforced piles and 12 cm for reinforced piles.

- Reinforcement: Strauss piles can be reinforced or not.

In the case of unreinforced piles subjected to compressive stress only, the reinforcement is just a starting reinforcement with no structural function and the steel bars can be placed in the concrete, one by one, without stirrups, immediately after concreting, leaving out the waiting time provided for in the design.

For reinforced piles subject to tensile or bending stresses, the reinforcement cage must be inserted into the casing before concreting. In this case, the socket must be smaller in diameter than the reinforcement. When not specified in the design, a minimum reinforcement consisting of at least 4 steel bars with a total length of 2.0 metres must be used.

The dimensions used for Strauss piles due to market availability and their respective working loads (or permissible structural loads) are shown below in Table 6.

Table 6: Dimensions usually used and respective workloads

Coating Tube Ø (cm)	Pile Ø (cm)	Workload (tf)	Outer Crown Ø (cm)	Probe Ø (cm)	Socket Ø (cm)	Wheelbase Between axles (cm)	Tablet
22	25	20	24	17	15	75	
28	32	30	29	22	20	96	Up to 15 metres
32	38	40	34	27	25	114	
35	42	50	37	32	30	126	
40	45	60	42	35	33	135	

Source: ABEF Manual (2002).

According to NBR 6122 (2010), piles with a spacing of less than 5 diameters should not be driven at intervals of less than 12 hours. This distance refers to the pile with the largest diameter. At least 1% of the piles, and at least one per site, should be exposed below the dragging level and, if possible, up to the water level, to check the integrity and quality of the shaft.

According to Falconi (1998), there are some advantages and disadvantages to having a Strauss pile.

- Advantages:

- No vibrations or shaking in neighbouring buildings;
- Possibility of executing the pile with the designed length;
- During drilling, it is possible to check for foreign bodies in the ground or boulders, allowing the location to be changed before concreting;
- It is possible to ascertain the different layers and nature of the soil, as taking samples allows comparison with percussion drilling;
- The equipment can be set up on small plots of land, and the piles can be driven where other types of equipment cannot;
- Autonomy, important in remote regions or locations; Relatively low cost; Easy to implement on ground above water level.

- Disadvantages:

- Difficult to execute below water level;

- Relatively small load capacity;

- Difficult to drive in resistant soils;

- In very soft saturated clays and submerged sands, where there is a high risk of the shaft splitting due to the ingress of soil (in these cases this solution is not recommended);

- Strict control of the pile's concreting is essential to ensure that there are no faults, as most accidents with these piles are due to deficiencies in the concreting (especially the discontinuity of the shaft) during the removal of the pipe, which completely invalidates the pile.

CHAPTER 3

METHODOLOGY

Initially, a bibliographical review was presented on the types of foundations most commonly used in civil construction, such as geotechnical soil investigations, shallow foundations and the concepts of all types of deep foundations. In order to provide a good practical and theoretical basis for the methodologies studied, this work was carried out by consulting academic and didactic books, technical standards (ABNT) and academic articles. Equipment manuals and catalogues were also used as a basis for the study.

After preparing the theoretical basis of the work, a case study was carried out in a residential building, where the entire Strauss pile construction process was followed, the problems of execution errors were discussed based on the prescriptions of the technical standards, and systematic and rational solutions for the construction process were presented. The data collection consisted of visual inspections and information from the structural plans provided by the engineer responsible for the work. After surveying the Strauss execution errors, a budget will be drawn up for the costs and materials spent on the foundation's crowning blocks and bucket beams, which were altered as a result of the execution errors. The results analysed point to a significant improvement in the productive and economic aspects of the execution stage, highlighting the importance of good supervision of Strauss pile execution.

CHAPTER 4

CASE STUDY

4.1 Project descriptions and location

The project where the study was carried out involves the construction of a Strauss pile foundation in a residential building, located at Rua José Antônio Franco, lote19ª - quadra B, Centro neighbourhood, Barbacena - MG. This residential building has three basements, a ground floor and five floors, as shown in figure 24. The entire building itself is made up of the land, 543.75 m², and the location is privileged, as it is located in the centre of Barbacena, factors that add even more value to this development.

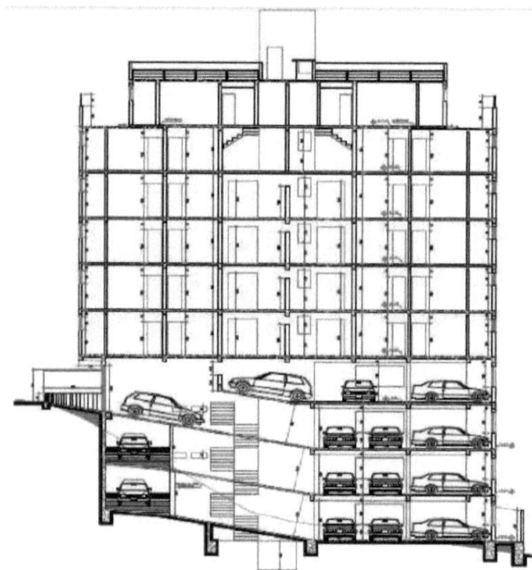

Figure 24: Sketch of the building section

Source: Author, 2017.

4.2 SPT surveys carried out

In order to identify the soil profile of the building under analysis, SPT borings were carried out on the ground between 11 and 13 March 2016, in two percussion boreholes (SP-01 and SP-02) totalling 27.75 m over an area of 543.75 m². In accordance with NBR 6484 (2001), simple reconnaissance boreholes were carried out. Contributing to the economic factor was the fact that these two boreholes were sufficient for a good soil investigation. The two boreholes carried out and the sketch of their location are shown in figure 25.

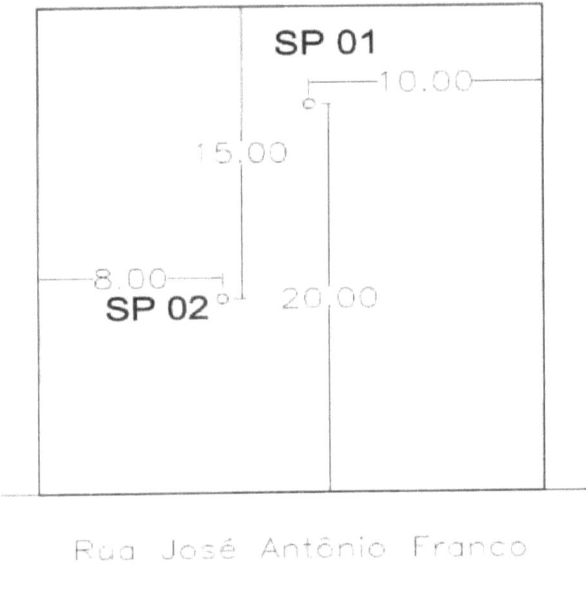

Figure 25: Borehole location plan
Source: GEO-BASE, Fundações LTDA, 2016.

The borings were carried out by percussion with the aid of water circulation and protected by a 2^" lining.

The test holes reached average depths of 13.8 m, with the water table at point SP 01 at 3.05 metres and at point SP 02 at 2.40 metres.

The soil was investigated and showed silty clay and sandy clay in the first layers, where it contained construction debris, as it was located on top of a landfill. In the layers below it showed variations with silty clay, clayey silt, with fine sand, sandy silt. Although the soil did not have good characteristics in terms of foundations, a more compact region was found capable of supporting the loads distributed by the piles. Figures 26 and 27 show the SPT results for boreholes 01 and 02.

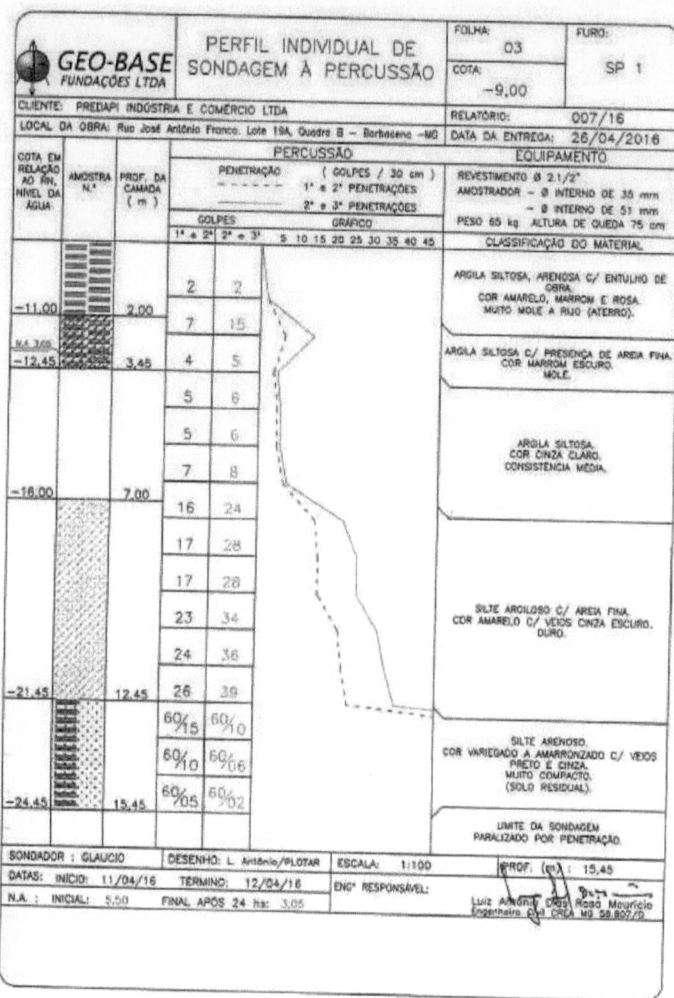

Figure 26: SPT test carried out in borehole 01
Source: GEO-BASE, Fundações LTDA, 2016.

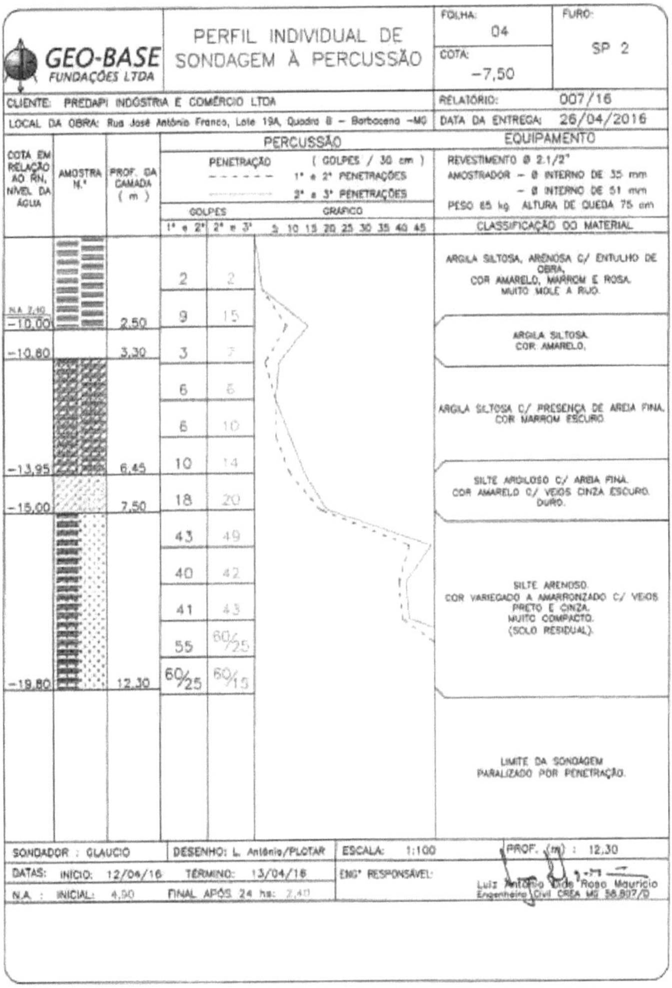

Figure 27: SPT test carried out in borehole 02
Source: GEO-BASE, Fundações LTDA, 2016.

4.3 Foundation type definitions

The Strauss pile used for the building's foundation was chosen on the basis of the following factors: soil type, neighbouring buildings (existence, minimising interference with neighbouring buildings), the possibility of installing the equipment in small areas or on sloping ground that is difficult for other equipment to access, and because it is relatively inexpensive. This stage was carried out by a specialised company, under the supervision of the engineer responsible for the execution and the foundation project.

A total of 63 piles were driven, 48 with a diameter of 52 cm and 15 with a diameter of 42 cm, and an average depth of 12.15 metres. Figure 28 shows the location of the piles.

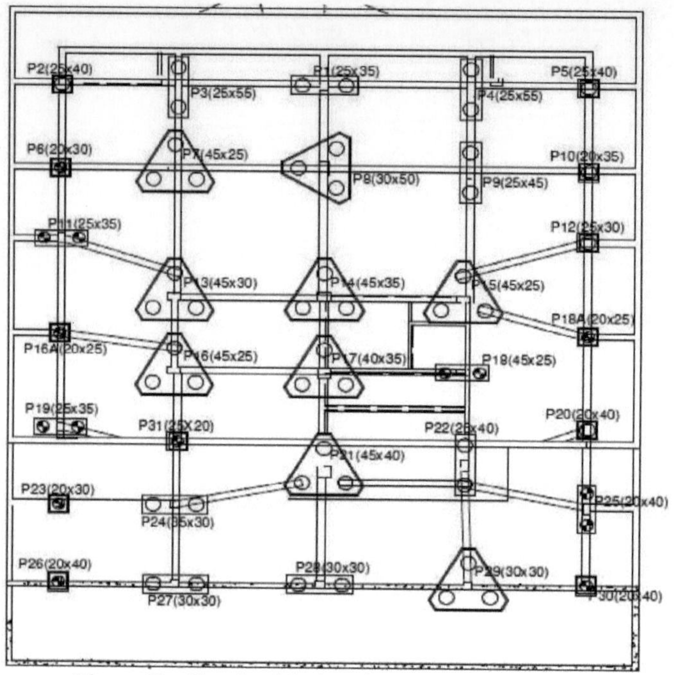

Figura 28: Sketch of the location plan for the piles and crowning blocks

Source: Author, 2017.

4.4 Preliminary services

Before starting to drill the piles, it is important to have some complementary documents at hand:

- Site survey report;
- Location plan with grading, reinforcement details and expected pile load;
- Executive report for the execution of each pile;
- NBR 6122:2010 - Design and execution of foundations - Procedures;
- NBR 6118:2014 - Design of concrete structures - Procedures.

4.5 Pile locations and pile driver positioning

Services are continued by locating the axis of the piles on the ground. Proper location is necessary to avoid possible eccentricities that can lead to incorrect execution, which can increase costs for the project. According to NBR 6122:2010, the maximum tolerance for eccentricity is 10% of the pile diameter.

4.6 Drilling

First, the piles to be drilled were marked, the pile driver was positioned on the ground and the mechanical probe was centred in the location picket. Releasing the probe to form a pre-borehole in the ground with a depth of 2 metres, as shown in figure 29.

Figura 29: Mechanical probe, pre-hole formation

Source: Author, 2017.

Next, the first lining tube with a toothed lower end, called the "crown", was placed, with the mechanical probe already inside, shown in figure 30.

Figura 30:   First tube being inserted

Source: Author, 2017.

The operator then manoeuvres the probe up and down, cutting the ground with the help of water thrown into the pipe, and the probe removes and discharges the excavated material through the longitudinal windows.

After crimping the first pipe, the next pipe was threaded on and the operation was repeated until the series of threaded pipes reached a resistant layer, as shown by the difficulty of advancing the crimping.

4.7 Concreting

The concrete was produced on site using CPIII class 32 cement, medium sand and gravels 1 and 2. Cement consumption was 300 Kg/m³ and fck = 25 MPa.

To carry out the concrete pour, the 220kg socket was placed, now with the function of tamping the concrete. Firstly, water is poured into the pipe to clean it, then the concrete is poured into the pipe to form the bulb, shown in figure 31.

Figura 31:   Socket ready to trim the bulb

Source: Author, 2017.

The shaft was then concreted, where the pipes were removed with a hand winch as the concrete was shovelled, until it reached approximately 30 cm above the pile's level, where the excess was removed to prepare the pile head.

4.8 Placing the reinforcement

Steel bars with stirrups were used for the reinforcement of the crowning blocks only. For each pile, 4 $\varnothing$= 12.5 mm bars were used, 3.0 metres long, 2.5 metres inside the pile and 0.5 metres to wait for the blocks, inserted while still in the fresh concrete, as shown in figure 32.

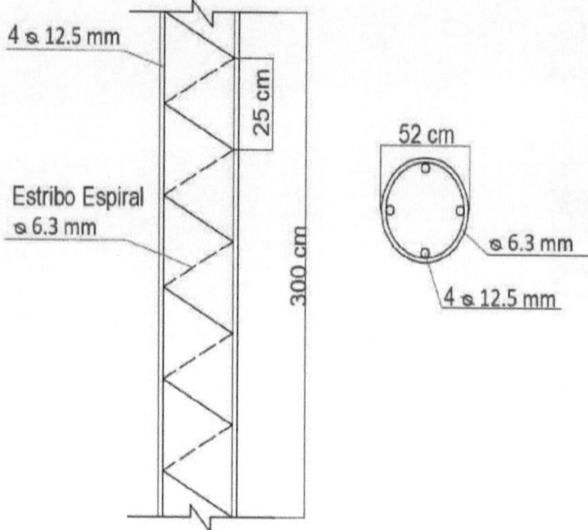

Figura 32: Pile reinforcement

Source: Author, 2017.

4.9 Execution errors

While monitoring the excavation work on the Strauss piles, some errors could be observed, and the lack of quality control in the execution was evident. At the end of the excavation, when the piles were ready and the equipment had been removed, the pillars were repositioned on the respective piles (according to the project), when they were found to be displaced, which affected both the safety and stability of the building.

4.9.1 Error in staking (P18)

In the piling of the pillar block (P18), according to the sketch of the pile location plan in figure 28, it can be seen from visual inspections that the pile drivers carried out two piles on the same day at an interval of 4 hours, where the two piles were spaced 150 cm apart, not respecting the curing time of the concrete from one pile to the next. According to NBR 6122 (2010), piles with a spacing of less than 5 diameters should not be driven at an interval of less than 12 hours.

If the piles are drilled in the same block, there could be major problems, and concrete could migrate from one pile to the next, where there would also be vibration in the pile that has already been concreted, due to the vibrations of the pile driver, which could cause cracks and damage them.

4.9.2 The pile is located out of position

This error resulted in eccentricity between the superstructure load and the foundation element. Several piles that were driven by the pile driving workers were not drilled into the pickets that fixed their markings in accordance with the design, resulting in several piles out of position, some with large errors and others with smaller ones. According to NBR 6122:2010, the maximum tolerance for eccentricity is 10% of the pile's diameter. In addition to the correct location, it is necessary to drive the piles with the minimum amount of displacement to avoid possible eccentricities, which leads to higher costs for the project. In the piling of the pillar block (P9), figure 33 shows the piles that had the greatest error in their location on site according to the planned project.

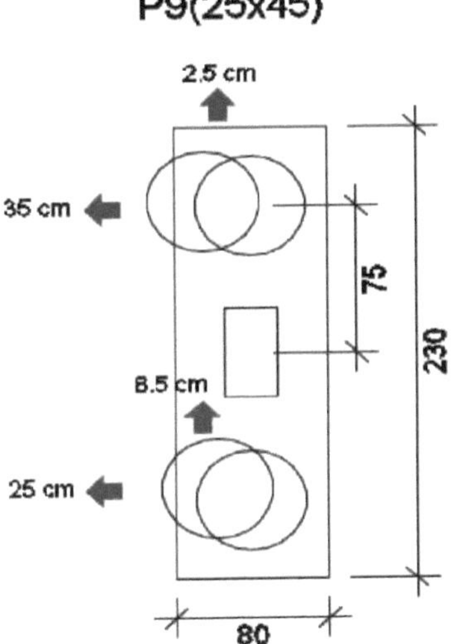

Figura 33: Column pile detailing (P9)

Source: Author, 2017.

Figure 34 shows the piles and anchors for the crowning block, where there was displacement.

Figure 34: Pile block of (P9)
Source: Author, 2017.

The eccentricity of the set of block piles in relation to the design location must then be compensated for with the use of beams to combat the forces (moments and shear), as well as promoting load adjustments on the piles. As a result, several baldram beams on the site had to be recalculated, including the baldram beam of the pillar 9 block (P9), which had a large increase in its section and reinforcement, as shown in the illustration of the beam stirrups in figure 35, which shows the beam stirrup as it was planned in the project, and then the beam stirrup as reinforced.

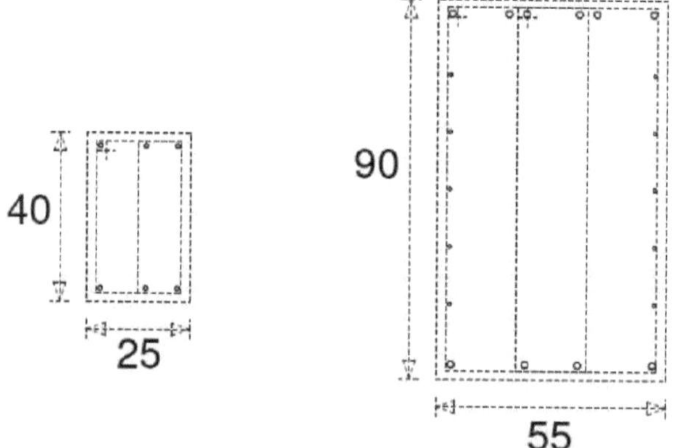

Figure 35: Stirrups before reinforcement and stirrups after reinforcement
Source: Author, 2017.

4.9.3 Failure to concrete the piles

Some piles had their final part compromised by contamination of the concrete by the surrounding material (soil), and did not have adequate resistance. In order to correct this problem, the pile head was pulled down to a slightly lower level, where the concrete was in adequate condition. Figure 36 shows the broken pile with low resistance at the top.

There must also be strict control of the concrete to be used in the piles, where the mix must be correctly executed as determined. And for this problem not to have occurred, concrete specimens should have been moulded for testing, which was not done.

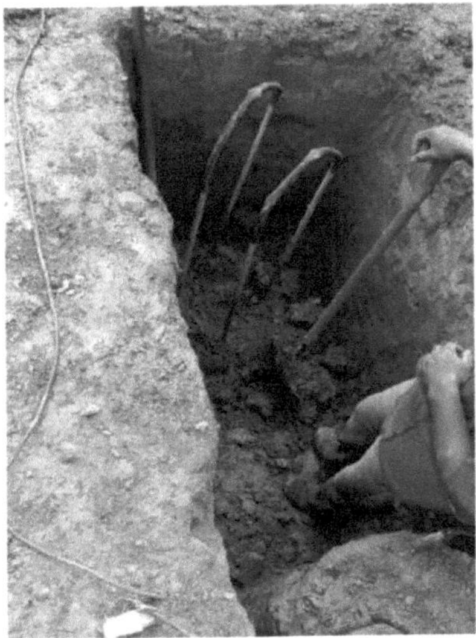

Figura 36: Pile with low-strength concrete

Source: Author, 2017.

Figure 37 shows the worker breaking the pile with a concrete breaker, until the pile's high strength is reached.

Figura 37:   Worker running hammer until pile strength is reached

Source: Author, 2017.

4.9.4 Damaged pile

There was excess water in the concrete of the pile at the end of the pillar (P2), and when the casing pipe was removed, the reinforcement dropped below its design level. This pile also had to be grouted more deeply.

In order to lengthen the foundation to the planned level, a crowning block was then built on the pile at its bearing level, along with two inclined blocking beams to give this block balance. From there, a pillar was started up to the level of the other blocks, and a new block was also built.

Figura 38: Damaged pile

Source: Author, 2017.

4.10 Drawing up the budget

In order to determine and analyse the costs of the work caused by the Strauss pile execution errors, the following options were adopted: the cost that the foundation would have had without the errors according to the planned project and the final cost of the foundation with the pile execution errors.

The budgets were based on the cost compositions found in the SINAPI table of the Caixa Econômica Federal and IBGE as at October 2017, with reference prices for the state of Minas Gerais, and on the SETOP spreadsheet with reference prices for building works in the state of Minas Gerais as at March 2017. The computer programme Microsoft Office Excel 2010 will be used to build spreadsheets for budgeting.

The budgets did not take into account the services and materials spent on the Strauss pile, but only the services and materials spent on the foundation's crowning blocks and beams.

BDI (Factor that represents the cost of additional expenses, profit and taxes) was taken into account in the proposed foundation budget, amounting to 25 per cent of the total.

The cost of executing the foundation without the execution errors was calculated at R$60,681.27. Table 7 shows a breakdown of the costs for executing the foundation in the planned project.

Table 7: Breakdown of services and materials for the foundation before errors

COMPOSITION	CODE	DESCRIPTION	UNI D	QUANT	P.UNIT [R$]	TOTAL P.A. (R$)
Setup	Ter-Esc-036	Manual trench digging	m³	77,38	41,58	R$ 3.217,46
SINAPI	74076/001	Plank formwork for concrete foundations reap 3x	m²	304,24	41,29	R$ 12.562,06
SINAPI	74254/002	Steel frameCA-50 supply, bending, cutting and fitting	kg	1130	7,31	R$8.260,30
SINAPI	73942/002	CA-60 steel frame supply, bending, cutting and fitting	kg	50	7,16	R$ 358,00
SINAPI	74138/002	Machined structural concrete fck= 25Mpa pouring, compacting and curing	Hl³	77,38	312,06	R$24.147,20
SUBTOTAL						R$48.545,02
BDI=25%						R$ 12.136,25
TOTAL						R$60.681,27

Source: Author, 2017.

The cost of executing the foundation with the execution errors was calculated at R$81,292.27. Table 8 shows the breakdown of the costs for executing the foundation with the costs of the errors that were made.

Table 8: Breakdown of services and materials for the execution of the foundation after the errors

COMPOSITION	CODE	DESCRIPTION	UNI D	QUANT	P.UNIT (R$)	TOTAL P.S. (R$)
Setop	Ter-Esc-036	Manual trench digging	m^3	92,67	41,58	R$3.853,21
SINAPI	74076/001	Plank formwork for concrete foundation reap 3x	m^3	390,65	41,29	R$ 16.129,93
SINAPI	74254/002	CA-50 steel frame supply, bending, cutting and fitting	-s	2152	7,31	R$ 15.731,12
SINAPI	73942/002	CA-60 steel frame supply, bending, cutting and fitting	kg	56	7,16	R$4OO,96
SINAPI	74138/002	Used structural concrete fck= 25Mpa casting, compacting and curing	m^3	92,67	312,06	R$ 28.918,60
SUBTOTAL						R$ 65.033,82
BDI= 25%						R$ 16.258,45
TOTAL						R$81.292,27

Source: Author, 2017.

When analysing the results found in the two budget options, it can be seen that the errors in the execution of the Strauss pile had a direct influence on the costs of executing the foundations for the project, with an increase of 34%, showing a loss.

CHAPTER 5

CONCLUSION

It can be concluded that this case study identified errors in the execution of the Strauss pile. It was evident that there was a lack of control in the execution, where the biggest problem was the lack of experience in the Strauss machinery, which was not executed according to the prescriptions of the technical standards, and in relation to the labour force, which was not qualified to execute the pile.

The importance of good supervision in the execution of Strauss is emphasised. This will prevent errors from appearing or correct them when they are at an early stage, preserving the original characteristics of the building and guaranteeing its safety and performance. Early detection of execution errors, through technical monitoring, is extremely important, because the sooner errors are observed, the lower the cost for the work.

Economically, it was observed that the poor execution of the Strauss pile presented much higher costs, compared to the errors in execution that were foreseen in the project, with an increase of 34%, showing a loss. More and more specialised workers are being sought in the execution of piles, with the guarantee and control of the final quality of the product, which points to a significant improvement in the productive and economic aspects, as well as the importance of encouraging further research on this subject that can encompass all the stages of the execution of a foundation, reinforcing the importance of compliance with the standards.

BIBLIOGRAPHICAL REFERENCES

ABNT - BRAZILIAN ASSOCIATION OF TECHNICAL STANDARDS. **NBR 6118: Design of concrete structures - Procedure**. Rio de Janeiro, 2014.

ABNT - BRAZILIAN ASSOCIATION OF TECHNICAL STANDARDS. **NBR 6122: Design and execution of foundations - Procedure**. Rio de Janeiro, 2010.

ABNT - BRAZILIAN ASSOCIATION OF TECHNICAL STANDARDS. **NBR 6484: Soil -**

Simple reconnaissance borings with SPT - Test method. Rio de Janeiro, 2001.

ALONSO, U.R. Execution of deep foundations: injected piles. In: HACHICH, W.; FALCONI, F.F.; SAES, J.L.; FROTA, R.G.Q.; CARVALHO, C.S.; NIYAMA, S. (Org.). **Fundações:** teoria e prática. 2. ed. São Paulo: Pini, 1998a . p. 361-372.

ALONSO, U.R. Execution of deep foundations: precast piles. In: HACHICH, W.; FALCONI, F.F.; SAES, J.L.; FROTA, R.G.Q.; CARVALHO, C.S.; NIYAMA, S. (Org.). **Fundações:** teoria e prática. 2. ed. São Paulo: Pini, 1998. p. 373399.

ALONSO, U.R.; GOLOMBEK, S. Execution of deep foundations: tubulars and caissons. In: HACHICH, W.; FALCONI, F.F.; SAES, J.L.; FROTA, R.G.Q.; CARVALHO, C.S.; NIYAMA, S. (Org.). **Fundações:** teoria e prática. 2.ed. São Paulo: Pini, 1998. p. 400-408. ANTUNES, W.R.; TAROZZO, H. Execution of deep foundations: continuous auger piles. In: HACHICH, W.; FALCONI, F.F.; SAES, J.L.; FROTA, R.G.Q.; CARVALHO, C.S.; NIYAMA, S. (Org.). **Fundações:** teoria e prática. 2. ed. São Paulo: Pini, 1998. p. 345-348.

BRAZILIAN PORTLAND CEMENT ASSOCIATION - ABCP. **ABCP Structures Manual**. São Paulo, 2002. Available at: <https://www.passei direto.com /arquivo/5793483/abcp-fundacoes> Accessed on: 07 July 2017.

BRAZILIAN ASSOCIATION OF FOUNDATION AND GEOTECHNICAL ENGINEERING COMPANIES - ABEF. **Manual for the execution of foundations and geotechnics**: recommended practices. 1. ed. São Paulo: PINI, 2012.

AZEREDO, Hélio. A. **The building up to its roof.** 2. ed. São Paulo: Edgard Blucher, 1997.

BUDHU, Muni. **Foundations and retaining structures**. 1. ed. Rio de Janeiro: LTC, 2015.

FALCONI, F.F. Execution of deep foundations: Strauss piles. In: HACHICH, W.; FALCONI, F.F.; SAES, J.L.; FROTA, R.G.Q.; CARVALHO, C.S.; NIYAMA, S. (Org.). **Fundações:** teoria e prática. 2. ed. São Paulo: Pini, 1998. p. 336-344.

MAIA, C.M. Execution of deep foundations: franki piles. In: HACHICH, W.; FALCONI, F.F.; SAES, J.L.; FROTA, R.G.Q.; CARVALHO, C.S.; NIYAMA, S. (Org.). **Fundações:** teoria e prática. 2. ed. São Paulo: Pini, 1998. p. 329-335.

QUARESMA, A.R.; DÉCOURT, L.; QUARESMA FILHO, A.R.; ALMEIDA, M.S.S.; DANZIGER, F. Geotechnical investigations. In: HACHICH, W.; FALCONI, F.F.; SAES, J.L.; FROTA, R.G.Q.; CARVALHO, C.S.; NIYAMA, S. (Org.). **Fundações:** teoria e prática. 2.ed. São Paulo: Pini, 1998. p. 119-162.

REBELLO, Yopanan C. P. **Fundações:** guia prático de projeto, execução e dimensionamento. 4. ed. São Paulo: Zigurate editoras, 2008.

TÉCHNE MAGAZINE. [São Paulo]: Pini, n. 83, feb. 2004. Monthly.

RODRIGUES, Edmundo. Study of foundations. [Rio de Janeiro]. Available at:<https://www.yumpu.com/pt/document/view/12937070/apostila-de-fundacoes-   ufrrj> Accessed on: 02 November 2017.

SAES, J.L. Execution of deep foundations: piles excavated with bentonite slurry. In: HACHICH, W.; FALCONI, F.F.; SAES, J.L.; FROTA, R.G.Q.; CARVALHO, C.S.; NIYAMA, S. (Org.). **Fundações:** teoria e prática. 2.ed. São Paulo: Pini, 1998. p. 348360.

SETOP-MG. **Reference Spreadsheet of Unit Prices for Building and Infrastructure Works.** Available at: http://www.setop.mg.gov.br /images/documents/precosetop/preco setop leste.pdf. Accessed on: 11 November 2017.

SINAPI. **National system for researching construction costs and indices (CAIXA).** Available at: http://www.caixa.gov.br/Downloads/sinapi-a-partir-jan- 2017-mg/SINAPI Custo ref Composicos MG 072015 Nao Desonerado.PDF. Accessed on: 03 October 2017.

TEIXEIRA, A.H.; GODOY, N.S. Analysis, design and execution of shallow foundations. In:

HACHICH, W.; FALCONI, F.F.; SAES, J.L.; FROTA, R.G.Q.; CARVALHO, C.S.; NIYAMA, S. (Org.). **Fundações:** teoria e prática. 2. ed. São Paulo: Pini, 1998. p. 227264.

VELLOSO, Dirceu de A.; LOPES, Francisco de R. **Foundations:** design criteria, subsoil investigation, shallow foundations, deep foundations. Complete volume. São Paulo: Oficina de Textos, 2010.

I want morebooks!

Buy your books fast and straightforward online - at one of world's fastest growing online book stores! Environmentally sound due to Print-on-Demand technologies.

Buy your books online at
www.morebooks.shop

Kaufen Sie Ihre Bücher schnell und unkompliziert online – auf einer der am schnellsten wachsenden Buchhandelsplattformen weltweit! Dank Print-On-Demand umwelt- und ressourcenschonend produzi ert.

Bücher schneller online kaufen
www.morebooks.shop

Printed by Books on Demand GmbH, Norderstedt / Germany